Preschool Science
Garden, Ocean and Weather
Themes for Content-Area Learning

Written by
Marie E. Cecchini
and Veronica Terrill

Illustrated by
Veronica Terrill and
Janet Armbrust

Teaching & Learning Company
1204 Buchanan St., P.O. Box 10
Carthage, IL 62321-0010

This book belongs to

Table of Contents

Cover design by Jenny Morgan and Sara King.

Copyright © 2007, Teaching & Learning Company

ISBN 13: 978-1-57310-528-6

ISBN 10: 1-57310-528-7

Printing No. 987654321

Teaching & Learning Company
1204 Buchanan St., P.O. Box 10
Carthage, IL 62321-0010

Several of the activities in this book involve preparing, tasting and sharing food items. We urge you to be aware of any food allergies or restrictions your students may have and to supervise these activities diligently. All food-related suggestions are identified with this allergy-alert symbol: ⚠

Please note: small food items (candies, raisins, cereal, etc.) can also pose a choking hazard.

At the time of publication, every effort was made to insure the accuracy of the information included in this book. However, we cannot guarantee that the agencies and organizations mentioned will continue to operate or to maintain these current locations.

Activities

Vegetable Sort 'n' Count

Reproduce, color and cut out the vegetable patterns on page 4, making several sets for each child. (Vary the numbers of vegetables in each set.) Have children sort and count their patterns by: color, type (above ground, vine, etc.) or number. Answers can be graphed.

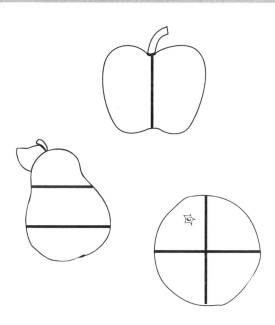

Fruit Fractions

Reproduce, color and cut out the fruit patterns on page 5. Mix up the patterns, and give one set to each child. Have children make whole fruits and identify: a half, a quarter and so on.

Sequence Sets

Reproduce the sequence sets on page 6. Cut the sections apart. Have children glue the sections together on an 8¹/₂" x11" sheet of paper in the proper order. Color the sheets when completed.

Vegetable Patterns

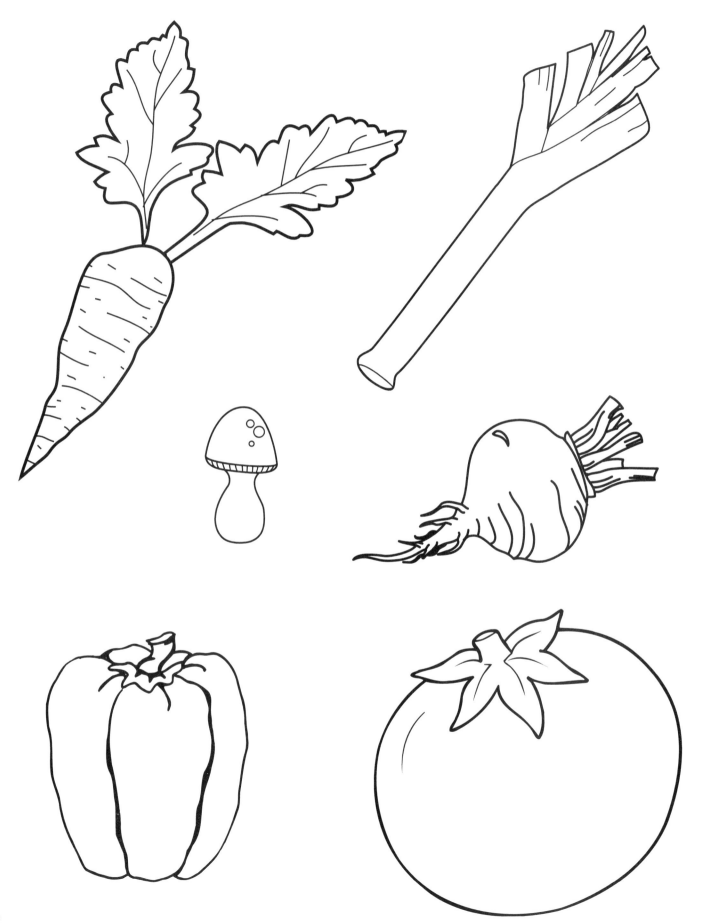

4

Fruit Patterns

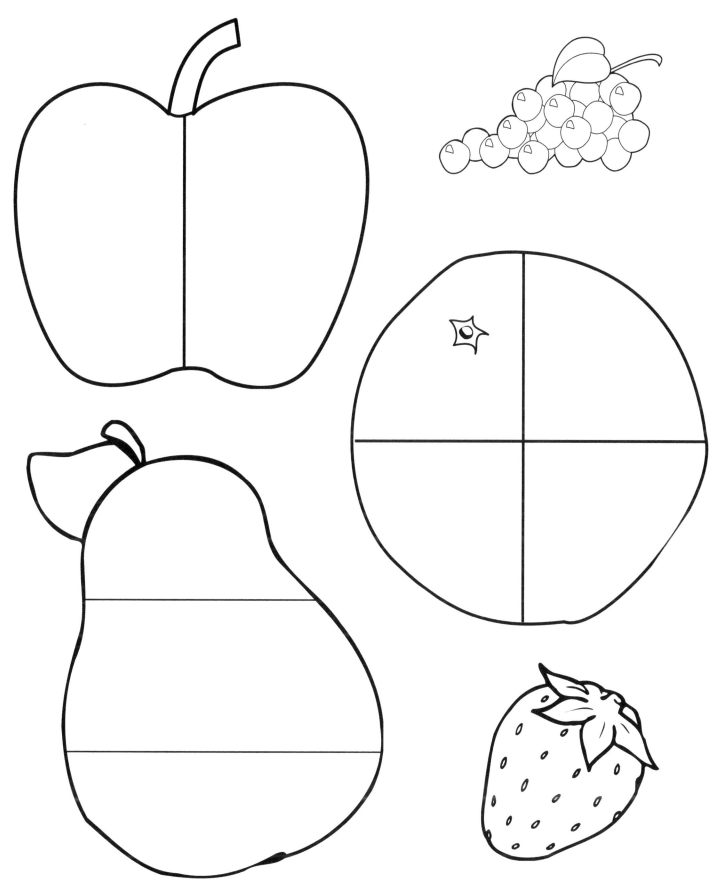

Note: These patterns may also be enlarged for room decorations.

Sequence Sets

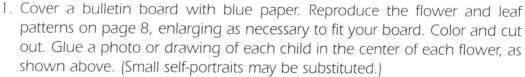

See How We've Grown!

1. Cover a bulletin board with blue paper. Reproduce the flower and leaf patterns on page 8, enlarging as necessary to fit your board. Color and cut out. Glue a photo or drawing of each child in the center of each flower, as shown above. (Small self-portraits may be substituted.)
2. On the bottom two leaves, glue writing samples from the beginning of the year, such as their name and numbers 1 - 5.
3. On the top two leaves, glue recent samples of the same words.
4. Attach the leaves as illustrated. Additional leaves can be added if desired.
5. Place the title *See How We've Grown!* across the top.

Bulletin Board Patterns

See How We've Grown!

1. Cover a bulletin board with blue paper. Reproduce the flower and leaf patterns on page 8, enlarging as necessary to fit your board. Color and cut out. Glue a photo or drawing of each child in the center of each flower, as shown above. (Small self-portraits may be substituted.)
2. On the bottom two leaves, glue writing samples from the beginning of the year, such as their name and numbers 1 - 5.
3. On the top two leaves, glue recent samples of the same words.
4. Attach the leaves as illustrated. Additional leaves can be added if desired.
5. Place the title *See How We've Grown!* across the top.

Bulletin Board Patterns

Books

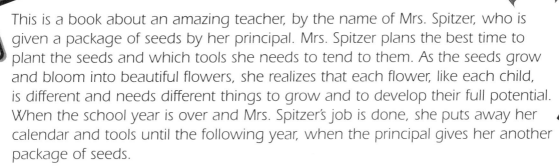

Mrs. Spitzer's Garden

by Edith Pattou, Harcourt, 2001

This is a book about an amazing teacher, by the name of Mrs. Spitzer, who is given a package of seeds by her principal. Mrs. Spitzer plans the best time to plant the seeds and which tools she needs to tend to them. As the seeds grow and bloom into beautiful flowers, she realizes that each flower, like each child, is different and needs different things to grow and to develop their full potential. When the school year is over and Mrs. Spitzer's job is done, she puts away her calendar and tools until the following year, when the principal gives her another package of seeds.

Jack's Garden

by Henry Cole, Harper, 1997

This book is an excellent step-by-step illustration of how a garden is planted and what the seeds and living creatures do after that. Readers are able to get a first-hand view of above and below the soil as the garden progresses throughout the season. The classic children's poem about the house that Jack built is modified into a catchy text for each picture, which makes understanding the magic of a little garden ecosystem effortless for kids as young as three years. A great book for ages 4 to 9.

Fluffy Grows a Garden

by Kate McMullan, Cartwheel Books, 2002

Ms. Day's class is planting a garden, and Fluffy is happy that Emma and Jasmine are planting carrots and peas. Fluffy's job is to guard the garden, which he takes very seriously. When some bees fly by, he tells them to buzz off! Then the bees tell him how they help his garden, and Fluffy decides to let them stay. He also meets some ladybugs and worms, also beneficial to his garden. But when he finds slugs, Fluffy sends them packing–they're not good for any garden!

9

Crafts

Super Sunflowers

Reproduce the sunflower pattern on page 11 onto heavy paper. Color and cut out. Glue sunflower seeds to fill the center, as shown. Attach the "sunflowers" high on a wall. Add green crepe-paper "stems" and "leaves."

Pretty Plant Stakes

Cut one $1/4$"-dowel to 18" for each child. Have children decorate the dowel with colorful, permanent markers or paints. Glue small sequins to the stakes. On top of each stake, wind a 2' length of wire several times around, and shape the ends. Attach bells or beads as illustrated.

Garden Pockets

Buy one cotton carpenter's apron for each child. Have children decorate their aprons using fruit and vegetable stamps. Cut the fruits and vegetables in half and dip into fabric paint. Press to make prints on the aprons. (These make great Mother's Day gifts!)

10

Sunflower and Leaf Patterns

Finish the Flower

Garden Fingerplay

We've got a little garden that we just adore,
(Children hug themselves and move from side-to-side.)

And we're growing lots of carrots in row number 4.
(Pretend to eat carrots and hold up 4 fingers.)

In our little garden, as you can plainly see,
(Hold hands above eyebrows and look around.)

We've got big, yellow sunflowers in row number 3.
(Reach both hands over head in a big circle, swaying—then put up three fingers.)

Now take a look in row number 2;
(Put hands up to eyes like pretend binoculars, then put up two fingers.)

We planted juicy watermelon, just for you!
(Pretend to hold a big, heavy watermelon, then point to each other.)

See us watering row number 1?
(Pretend to water the ground and put up one finger.)

Our gardening fun has just begun!
(Throw arms out wide and smile great big!)

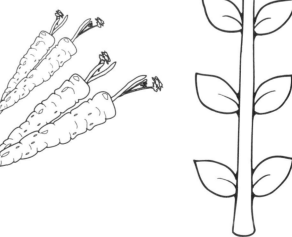

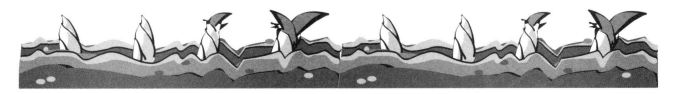

My Garden

On a cool April day I get my shovel out
And go out to dig a garden spot, to move the dirt about.

It's chilly, but I soon warm up from digging in the dirt,
So I stop to roll up the sleeves of my shirt.

My happy cocker spaniel is digging too, I see.
I wonder, does he really think that he is helping me?

Two weeks later it is time to plant my garden seeds.
I gather seeds and strings and sticks, and other things I'll need.

I stretch a string along each row and tie two small sticks to it,
Then plant my seeds in long, straight lines, 'cause that's the way to do it!

Tomato seeds and radishes, cucumbers and zucchini—
How can such big vegetables come from seeds so teeny?

Watermelon, green beans and corn on the cob
Will sprout and start to grow, and soon I'll have another job.

Sometime in mid-summer I'll have a lovely crop,
And I'll pick veggies till I fill my basket to the top.

Then I'll take them in the house and wash them carefully,
And eat the yummy vegetables planted by ME!

by Mary Tucker

Snacks

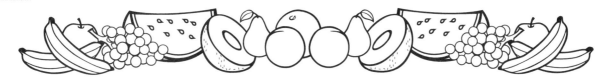

Dirt and Bugs

Fill small plastic cups with alternating layers of chocolate pudding and crushed chocolate sandwich cookies. Top with several candy worms.

Eat a Flower

Give children plates of cut vegetables, such as carrot rounds, celery sticks, etc. Let children create their own "flower" by assembling the vegetable pieces on a paper plate. Don't forget some chive "grass" or radish "roses."

Make Your Own Garden Sandwich

Create a sandwich bar of fresh vegetables and a variety of rolls and breads. Include spreads, such as veggie cream cheese and ranch dressing. Let children make their own garden sandwiches. How many different vegetables can they put in their "gardens"?

⚠ Make sure you are aware of any food allergies or restrictions children may have.
Be sure children wash hands before they eat.

My Garden Book

_____

Name

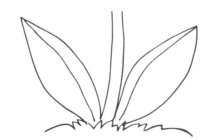

Can you draw three flowers?

1

Plants need lots of water!

Can you draw 12 more drops of water for the garden?

2

Plants need sunshine too! Draw a big sun in the sky.

3

It's harvesttime! Draw your favorite vegetables in the basket.

4

Gardening is great fun!

5

Colorful Fun

Color each picture the correct color.

red

green

pink

yellow

blue

brown

Count and Color

Circle the number that shows how many of each.

1 2 3 4 5

1 2 3 4 5

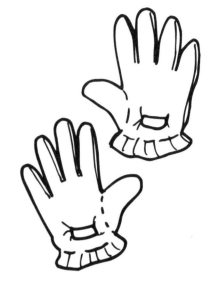

1 2 3 4 5

1 2 3 4 5

1 2 3 4 5

1 2 3 4 5

20

TLC10528 Copyright © Teaching & Learning Company, Carthage, IL 62321-0010

Number Skills

Count the objects in each set. Circle the number that shows how many are in each set.

1 2 3 4 5

1 2 3 4 5

1 2 3 4 5

1 2 3 4 5

1 2 3 4 5

1 2 3 4 5

Dot-to-Dot

Follow the numbers to see what's happening in the garden.

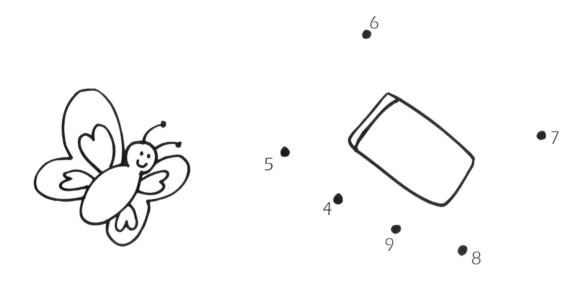

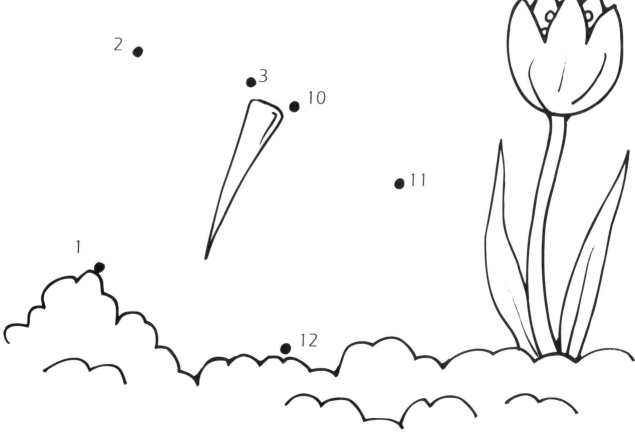

Name _____

Matching Sounds

Look at the pictures. Think of the letter that makes the first sound of each.
Draw a line from the picture to the correct sound.

s

g

p

a

f

Garden Maze

Can you make sure the flower gets some water?

Name _____

My Garden Page

25

See you
in the
Garden!

Seeds

Clip Art

Book Cover

All
About
Gardens . . .

_____
Name

Name _____

Garden Poster to Color

Activities

Science Activities

Buoyancy

Demonstrate how the salt in the ocean helps sea animals float. You will need two clear glasses, water, salt, two fresh eggs and a spoon. Pour water into both glasses. Add salt by the spoonful to one of the glasses and stir to dissolve. Continue to add salt until it will no longer dissolve. Break open an egg to represent the sea animal and pour the contents of the shell into the fresh, unsalted water. Pour the contents of the second egg into the salt water. What happens? Which kind of water makes it easier to float?

Which Is Heavier?

Find out which is heavier, fresh or salt water. You will need a clean empty mayonnaise jar, water, salt, a fresh egg and a spoon. Fill the jar halfway with water. Again, add salt by the spoonful and stir until the salt will no longer dissolve. Place the egg in the jar. What happens? The egg floats in the salt water. Now, slowly pour fresh water over the egg. What happens? The egg continues to float above the salt water, but is now below the fresh water. Help children conclude that the salt water is heavier than the egg, so the egg floats. The egg is heavier than the fresh water, so it remains at the bottom in the fresh water. Which is heavier, fresh or salt water?

Reflections

Display color photographs of oceans in various parts of the world. (Check out local libraries and travel agencies for pictures.) Discuss the pictures with children. What color is the water in each? Why does the water appear to be different colors? Experiment to find out where these colors come from. You will need a clear glass jar filled with water; a mirror; construction paper in blue, green, gray and black. Set the jar of water on the mirror. Hold the different colors of paper over the jar. What happens to the water color when you change the paper color? Why? Discuss how water reflects the colors of things like animals, plants and the sky.

Math Activities

Estimate and Count

Fill two small, clean fishbowls with goldfish-shaped crackers. Use small fish shapes for the first bowl, large fish shapes for the second. Challenge children to estimate how many fish are in each bowl. Empty the fishbowls one at a time, counting to find out how many fish were in each bowl. Whose estimate was nearest the actual amount?

Answer Fish

Draw and cut fish-shaped cards from construction paper or tagboard. On half of the cards, write a number from 1 to 20. On the second half of the cards, draw dots in configurations from 1 to 20. Mix the cards, then lay them facedown on a tabletop or the floor for a game that can be played by two or more children. Each player takes a turn to flip over two cards. If the number card matches the dot card, the player gets to keep both and take a second turn. If the cards do not match, they are returned to the facedown position and it is the next player's turn.

Graphing Preferences

Provide children with various samples of seafood, such as tuna, crab or anchovies to taste. (Make sure none of the children have seafood allergies.) Prepare a piece of poster board with the title *See Our Seafood* and a column for each seafood you taste. Label each column with the name of one seafood. Have children place an ocean sticker in the column of his or her food preference. Which food was the most/least favorite of the class?

⚠ Make sure you are aware of any food allergies or restrictions children may have. Be sure children wash hands before they eat.

Language Activities

Sounds

Cut out a large fish shape from poster board or construction paper. Have children search through old magazines and catalogs for pictures of items whose names begin with *F.* Glue these pictures to the large fish.

Variation: Have children name words that rhyme with *fish.* Write these rhyming words on the large fish.

Letters

Cut fish shapes from tagboard or construction paper. (Or use pre-cut fish shapes.) Write a different letter of the alphabet on each fish. Let children use the fish to spell their names and other words they know. Challenge them to place the fish in alphabetical order.

Extension: Cut out several ocean pictures from clip art. Work with children to name the animals and place the pictures in alphabetical order.

Sharing Ideas

Display and have children name pictures of several common undersea animals (octopus, eel, starfish, lobster, dolphin, etc.). Ask individual children to choose an animal they would like to be. Have them tell why they would like to be this animal, naming characteristics and behaviors of the animal that appeal to them. Ask individuals to name an animal they would definitely not want to be. Explain why.

Imagine

Provide children with paper and markers or crayons. Challenge them to draw and color a picture of a new sea creature, one that they have invented themselves. Have them name their creations, then write or dictate "facts" about that animal, such as what it likes to eat. Add the words under the pictures.

Label and Learn

Work with children to create a list of the many foods we eat that come from the ocean. Invite parents and/or staff members from different cultures to talk about and share samples of foods of the sea that they eat.

⚠ Make sure you are aware of any food allergies or restrictions children may have.
Be sure children wash hands before they eat.

1. Cover the board with dark blue paper. Drape several rows of light blue crepe-paper streamers across the board to represent water. Add pre-cut or self-made letters at the top to read *Ocean View*.

2. Crunch strands of green crepe-paper streamers to make seaweed stems. Add these to the board. Let children draw and cut green construction paper leaves to add to the seaweed.

3. Let children use paper in neon colors with neon markers to draw and color sea animals, such as dolphins, starfish, sharks, lobsters, crabs, sea horses and so on. Have children cut out their creations and position their work on the board.

Books

Alphabet Sea

by Carolyn Spencer and David Harris, Tortuga Books, 1999

A photographic alphabetical journey through tropical waters. Rhyming text and colorful photos help younger students learn the alphabet and invite slightly older students to practice basic reading skills.

First Hidden Pictures: Ocean

by Judith Jango-Cohen, McClanahan Book Co., 2000

Children learn to recognize sea creatures as you search for them in the colorful illustrations.

Hello, Ocean

by Pam Munoz Ryan, Tailwinds, 2001

A rhyming picturesque trip to the sea experienced through a young girl's five senses. Challenge children to match actual seaside finds (shells, sand dollars, etc.) with pictures of the same items in the book.

The Ocean Alphabet Book

by Jerry Pallotta, Charlesbridge, 1990

Take an alphabetical journey through the north Atlantic Ocean. Youngsters will learn fascinating facts about ocean life through vivid pictures and interesting topics. A great introduction to life under the sea, this book lends itself to various extension activities.

Somewhere in the Ocean

by Jennifer Ward and T.J. Matsh, Rising Moon, 2000

A rhyming number, song-story as told by sea animal families. Introduces many new words and invites further exploration. Contains a glossary and map to enhance classroom use.

This Is the Ocean

by Kersten Hamilton, Boyds Mills Press, 2001

Colorful pictures and rhyming text are used to describe the part the ocean plays in the water cycle.

Crafts

Fish Net Pictures

Materials: light blue construction paper, plastic mesh berry baskets, brown tempera paint, foam or aluminum paint trays, fish-shaped crackers, glue, scissors

Directions

1. Pour a little brown paint into the paint trays. Have children dip the bottom of a berry basket into the paint and use the basket as a stamp to create fish netting on the blue paper. Allow paint to dry.
2. Cut the papers into irregular shapes to resemble fish nets. Glue fish crackers onto these nets.

Sea in a Bottle

Materials: 1-liter soda or water bottles, water, blue food coloring, small plastic plants and fish, silver glitter

Directions

Have children place a few plastic plants and fish into their bottles. Sprinkle in a little glitter. Fill the bottles with water, add a few drops of blue food coloring, cap tightly and then shake to make the sea come alive.

Sea Bands

Materials: strips of construction paper for making headbands, small clip art ocean-themed pictures or stickers, markers, glue, stapler, scissors

Directions

1. Have children color and cut out clip art ocean pictures and glue these on their paper strips. Children can also fill the strips with seashore stickers.
2. Fit the paper strips around children's head. Staple the ends of the strips to make seaside headbands.

Giant Stuffed Fish

Materials: roll of craft or butcher paper, tempera paint, brushes, scissors, newspapers, stapler, yarn, hole punch

Directions

1. Work with small groups of children to draw and cut identical fish shapes from craft or butcher paper. Have children paint one side of each fish shape.

   Note: Each pair of fish cut-outs will be put together to make one fish. Be sure children paint the shapes accordingly.

2. When the paint dries, staple each pair of fish shapes together around the edges. Leave an opening for stuffing. Have children crumple newspapers and gently stuff their fish. Staple the opening in each fish closed.

3. Punch two holes at the top of each fish. Use yarn to suspend the fish from the classroom ceiling.

Starfish Puppets

Materials: foam trays, sand, yarn, pencils, scissors, glue, hole punch

Directions

1. Have children draw and cut starfish shapes from the foam trays. Punch a hole at one end of each starfish shape.
2. Spread glue over the starfish and sprinkle the glue with sand. Shake off excess sand and allow glue to dry.
3. Thread and knot one end of a length of yarn through the hole in the starfish. Tie the opposite end of the yarn into a loop for holding. Then hold the loop and gently pull the starfish across the floor to make the puppet move.

36

Poems

Silly Verses

If I were an octopus,
Do you know what I'd do?
I'd drive down to the mall,
And buy four pairs of shoes.

If I were dolphin,
Do you know where I'd swim?
I'd get myself a card
For the pool at the gym.

If I were a jellyfish,
Do you know where I'd float?
In a giant bubble bath,
And play with plastic boats.

If I were a hermit crab,
Do you know where I'd hide?
I'd climb onto a Ferris wheel,
And have myself a ride.

And if I were a lobster,
Do you know where I'd be?
Laying on my couch,
Just watching TV.

Fish School

Five little fish went off to school,
And the 1st one learned the Golden Rule.
The 2nd one learned to write her name,
And the 3rd one learned to play a new game.
The 4th one learned his ABCs,
And the 5th one learned her 1, 2, 3s.
The five little fish thought they were pretty smart,
And their teacher said, "You're off to a good start."

Use the poem as a fingerplay, having children point to their first, second, etc., fingers as they recite the rhyme. Talk about what children have learned in school. Compare this with what the fish learned. Ask children to tell what the teacher meant when she told the fish they were "off to a good start."

Games

cean

Sea Sponge Game

Equipment
supply of sea sponges (available at craft and hardware stores), two buckets, tape

To Play
Divide class into two teams. Members of each team will take turns tossing three to five sponges at their respective buckets. Record the number of sponges that went into each bucket. Add up the number of each player's sponges to find out which team got the most sponges in their bucket.

Variation
1. Toss sponges through a hoop suspended from the ceiling.
2. Make a bull's-eye target from craft paper. Tape it to the floor. Let children toss the sponges at the target to try and get a bull's-eye.

Extensions
- Use the sponges to play catch with a partner.
- See how many times individual children can toss a sponge in the air and catch it before it hits the floor.

Ping-Pong Fishing

Equipment
tub of water (a water table may also be used), several Ping-Pong™ balls, permanent marker, aquarium fish-catching nets (available at pet stores)

Setup
Use permanent marker to write a different number or letter on each Ping-Pong™ ball. Float the balls in a tub or table of water.

To Play
The number of players will be determined by the number of fish nets available. Each child will use the fish net to scoop a ball from the water. The child will read the name of the number or letter on the ball, then return the ball to the water. Change players after each has had five turns.

Snacks

Pizza Fish

sliced bread   tomato sauce
cheese slices   fish-shaped cookie cutter

Toast a slice of bread for each child. Place a small amount of room-temperature tomato sauce on the toasted bread slices. Cut cheese slices into fish shapes with the cookie cutter. Place the fish on top of the tomato sauce. Place bread slices under the broiler or in a warm oven just long enough for the cheese to melt.

Juice Fish

3 T. gelatin   cooking spray
2 c. boiling water   oblong baking dish
fish-shaped cookie cutter
12-oz. can frozen juice concentrate, thawed

Pour thawed juice concentrate into a mixing bowl. Stir in gelatin. Carefully add boiling water and stir to dissolve gelatin. Lightly grease baking dish with cooking spray. Pour juice mixture into baking dish. Refrigerate until firm. Using the cookie cutter, cut into fish shapes to eat.

Sand Cones

2 4-oz. packages instant banana pudding   4 c. milk
vanilla wafers   ice-cream cones

Prepare pudding as directed on the package. Place vanilla wafers in a self-seal plastic food storage bag. Crush wafers with a rolling pin or hammer. Stir crushed wafers into the pudding. Refrigerate mixture until pudding is set. Spoon into ice-cream cones to serve.

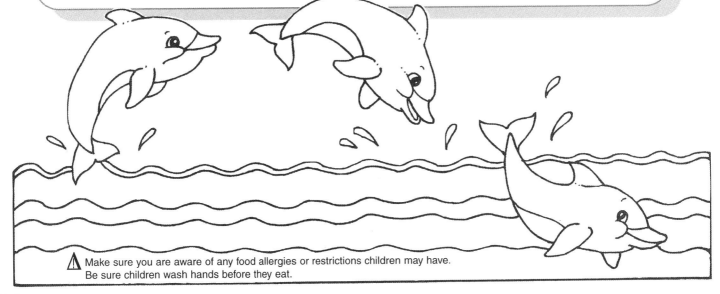

⚠ Make sure you are aware of any food allergies or restrictions children may have. Be sure children wash hands before they eat.

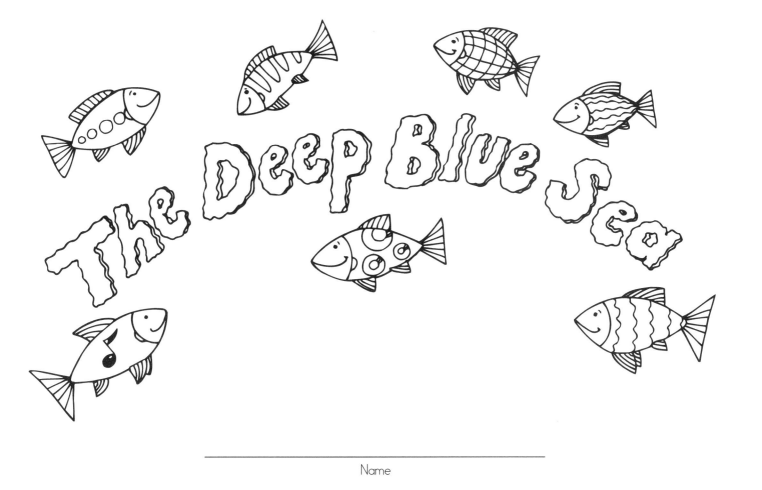

The Deep Blue Sea

_____

Name

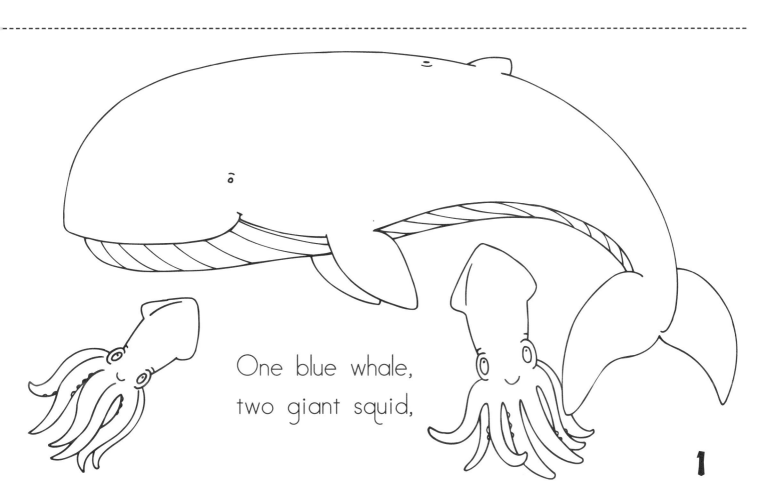

One blue whale,
two giant squid,

1

three jellyfish,
four sea turtles,

2

five flying fish,
six gray dolphins,

3

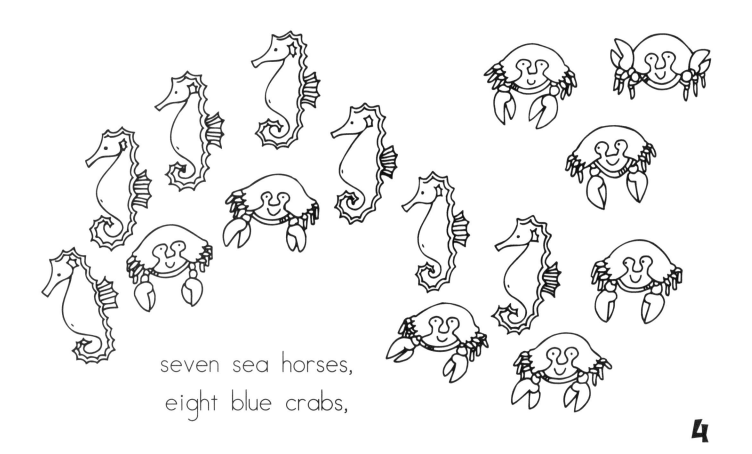

seven sea horses,
eight blue crabs,

4

nine spiny starfish

5

and ten round clams,

6

all live together
in the deep blue sea.

7

Sea Words

Draw a line to connect each picture with the correct word.

   **fish**

   **clam**

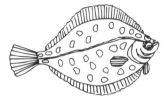   **lobster**

   **starfish**

   **octopus**

   **eel**

What Can It Be?

Follow the dots from a to z
to find out what is lurking beneath the sea.

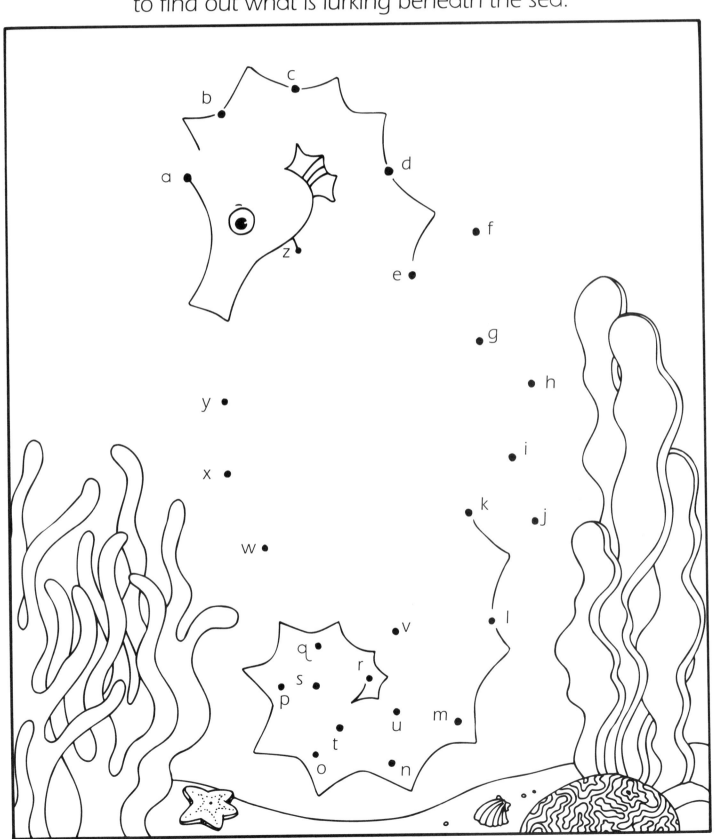

Undersea Detectives

Find and color 10 sea animals hidden in the seaweed.

Count and Match

Count the number of spots on each fish.
Draw a line to connect each fish with the correct number.

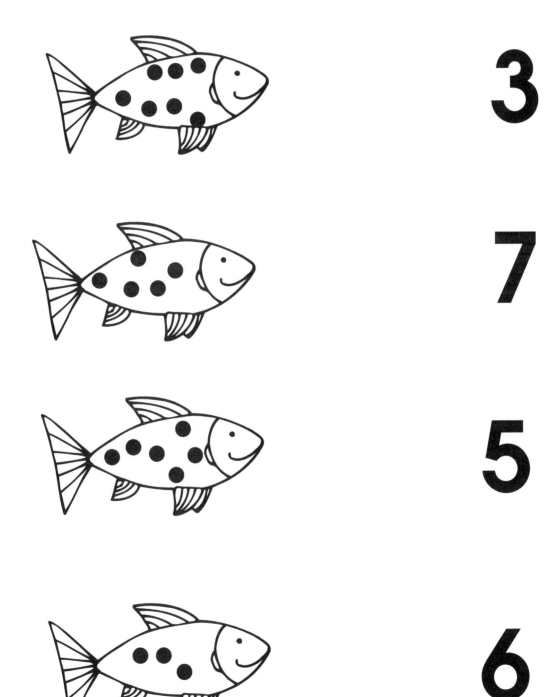

3

7

5

6

Fish School

Count the number of fish in each school.
Draw a line to connect the matching school of fish.

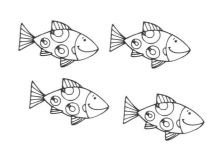

Large and Small

Color the fish. Cut them out. Glue them to another sheet of paper in order of size: from largest to smallest.

Name

Ocean Match

Color the matching sea creatures the same color.

Fish Puzzle

Color the fish. Cut it out. Cut the fish apart along the broken lines to make a puzzle. Try to put your puzzle back together by yourself.

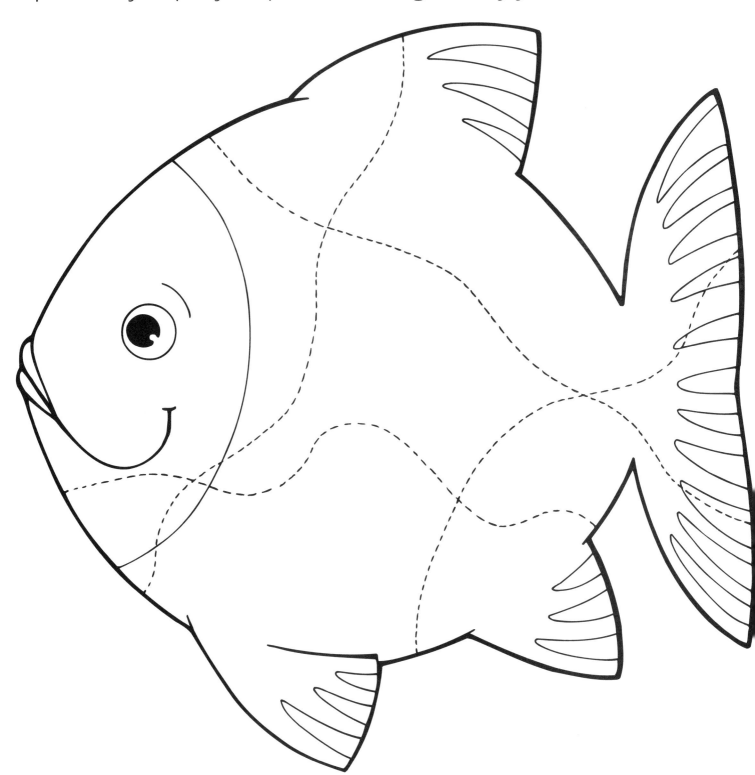

52

Sea Turtle Maze

Help the baby sea turtle find its mother.

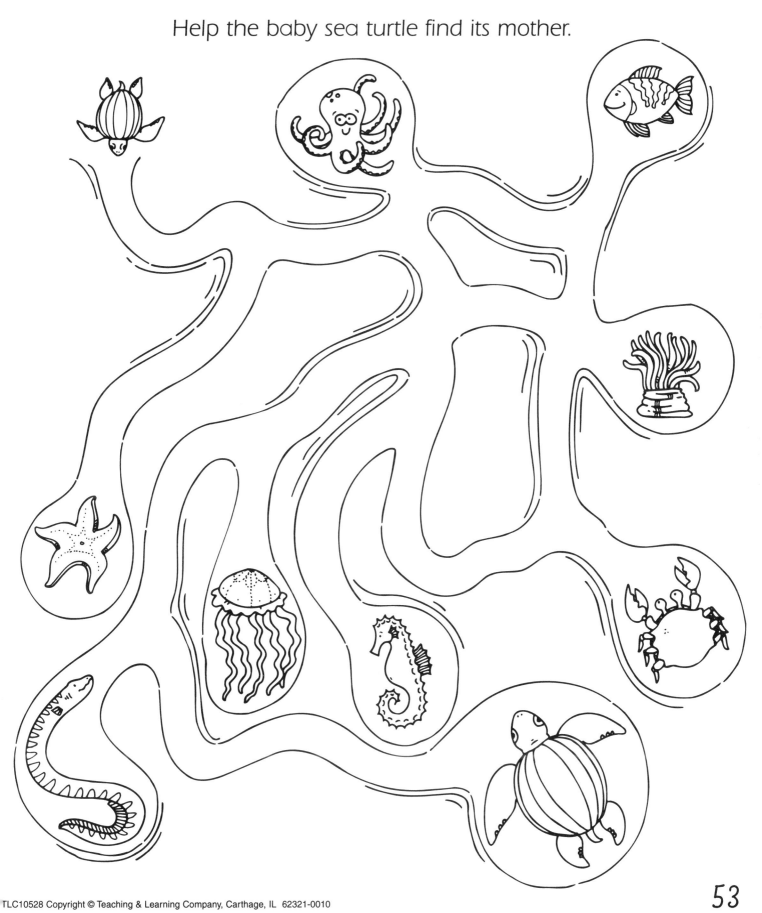

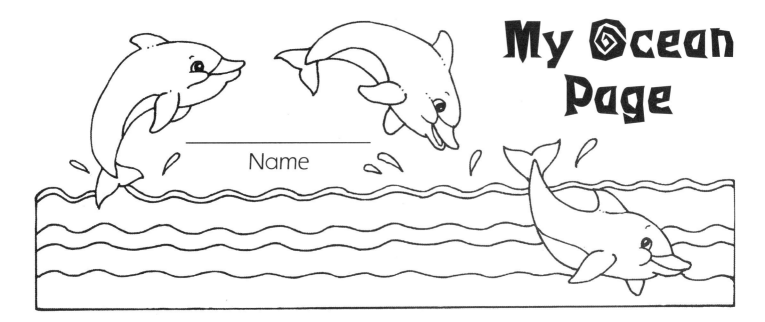

My Ocean Page

_____
Name

54

Clip Art

Clip Art

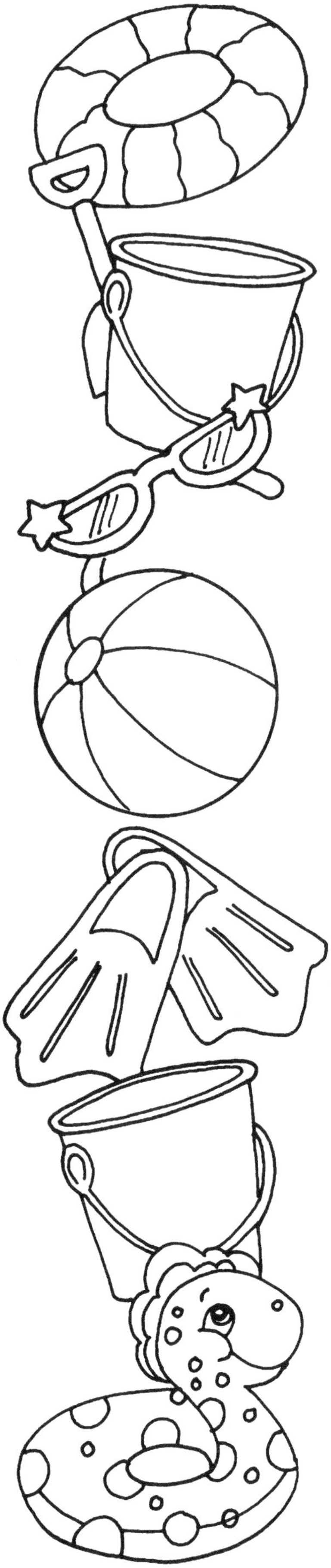

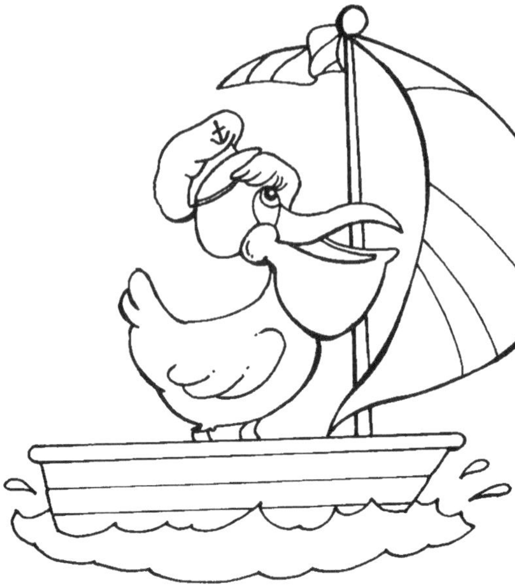

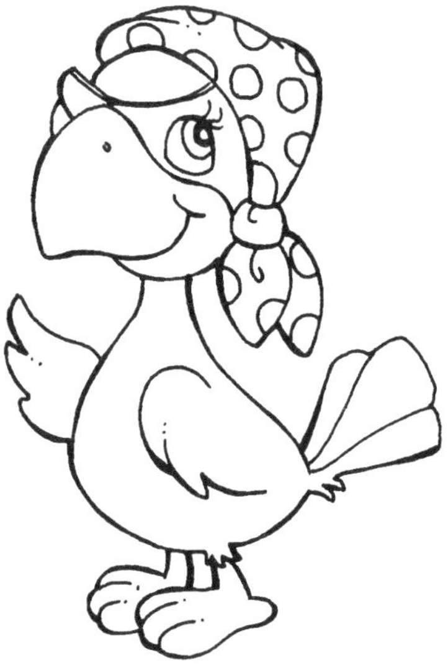

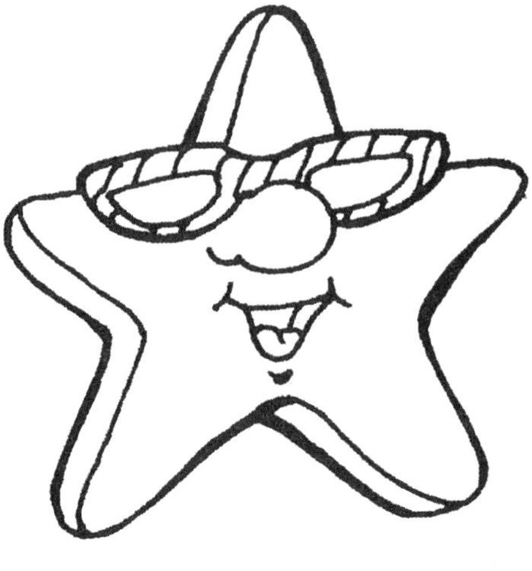

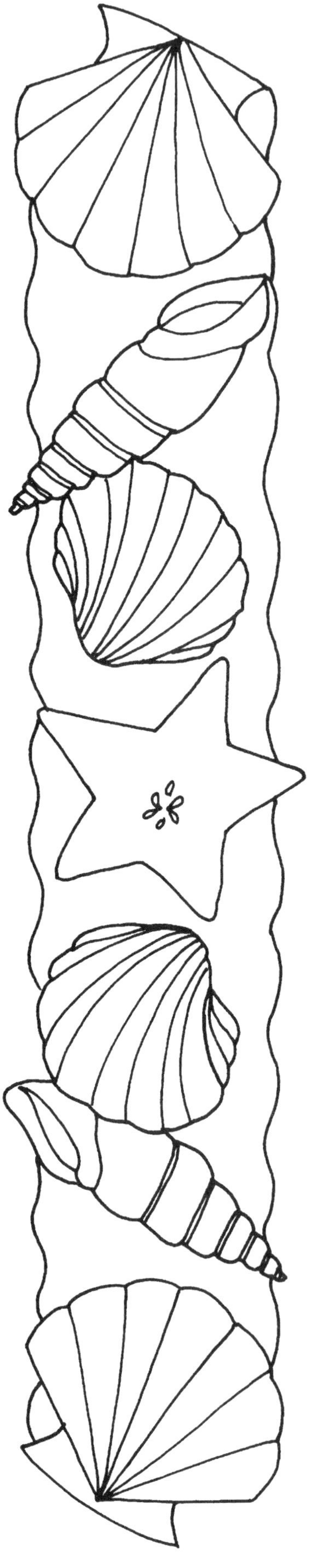

Book Cover

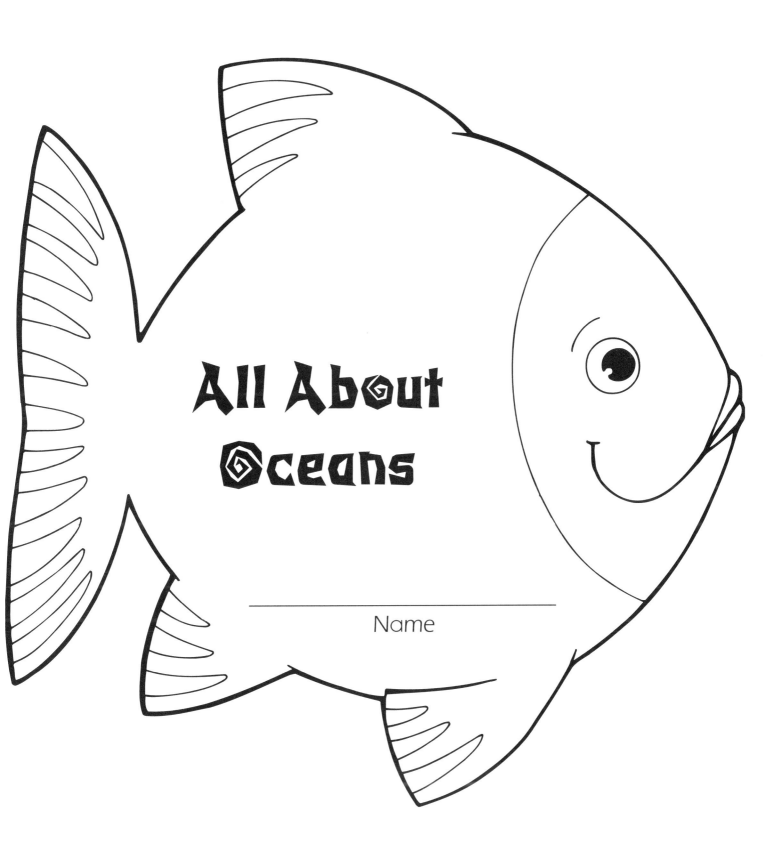

All About Oceans

_____
Name

Yarn Temperature Graph

On a blank wall or bulletin board, measure out a "thermometer" on the left-hand side of the area, basing the degrees depending on the season of the year. Across the top of the area, make a label for each day of the week. Cut a piece of colorful yarn to mark the high temperature of the previous day. The low temperature may also be marked with a colored pushpin or piece of tape. Add an appropriate piece of weather clip art (pages 77-78) at the top of each piece of string.

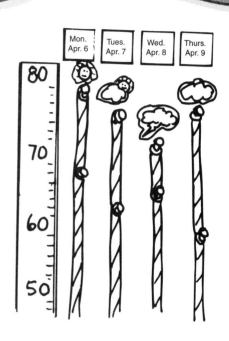

Go Fly a Kite

From an inexpensive kit or easy pattern, assemble a kite for each child or small group. Then locate a large area that is free of power lines or other obstacles. For several days in a row (depending on the weather, of course!), fly the kites and record your observations on a large chart. For example, "flew easily," "wouldn't fly," "kept dropping" and so on. Also record other indications of the wind such as "flag was fully out," "our hair was blowing" or "the leaves were rustling." Use this chart to illustrate a weather lesson on the "invisible" wind and its properties.

Weather Opposites

Cut several large blank index cards in half, lengthwise. Cut the strips in half again, making "puzzle-shaped" cuts, different for each set. On matching sets, letter each half with an "opposite" weather word or picture, such as *sunny* and *cloudy* or *raincoat* and *sunglasses*. (Small stickers may also be used.) Mix the pieces up and give to children to make matches. Store the pieces in a large envelope decorated with clip art from pages 77-78.

58

Bulletin Board

Rain or Shine?

APRIL

1. Cover a bulletin board with blue paper. Reproduce the cloud and sun patterns on page 60, enlarging as necessary to fit your board. Color and cut out. Position pieces on the board as shown.

2. Divide the board in half with a black line and add the heading *Rain* (or *Snow* depending on the time of year) or *Shine?* Also add the name of the month. Then thumbtack individual cards with the days of the month across the bottom of the board, as illustrated.

3. Have children watch the weather each day, at a designated time, and place that day's card on the appropriate side of the board, either *Rain* or *Shine*.

4. At the end of each month, you can graph the results to extend the lesson.

Bulletin Board Patterns

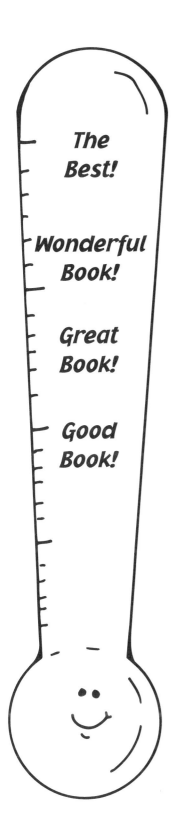

The
Best!

Wonderful
Book!

Great
Book!

Good
Book!

Kicks Up a Storm: A Book About Weather (Magic School Bus)
by Nancy White, Scholastic, 2000

When Ralphie imagines he's a superhero named Weatherman, the Magic School Bus becomes a glider riding an updraft into the storm clouds. Then the kids become part of the storm—turning from ice crystals to rain. A science lesson they'll never forget!

What Will the Weather Be?
by Lynda Dewitt, Scott Foresman, 1993

This book delves into the reasons why the weather is so hard to predict. Even though meteorologists know more than ever before, the changes outside just aren't always predictable. This book explains what scientists do know and what they sometimes can't know. This is a great supplemental tool for a weather unit.

Weather Words and What They Mean
by Gail Gibbons, Holiday House, 1992

This is a great book to use when teaching beginning weather. The words and illustrations are at a perfect level for young children.

Flash, Crash, Rumble, and Roll
by Franklyn Mansfield Branley, HarperCollins, 1999

Storms can be very scary to young children, but this book helps by explaining what causes storms. Great facts (lightning bolts can be over a mile long) and simple experiments (how to make a cloud) make this one of the best weather books around.

Sing Along with the Weather Dude
by Nick Walker, Small Gate Media, 2000

A book/CD combination that will wow everyone who listens. Catchy tunes make this a perfect way to learn more about weather. Contains a wealth of educational information.

Crafts

Fancy Fan

Decorate an 8½" x 11" sheet of construction paper, using crayons or markers. Fold one short side in, about 1". Fold the new edge the opposite way. Repeat until the whole sheet is folded. Gather and staple one end. Tie with ribbons.

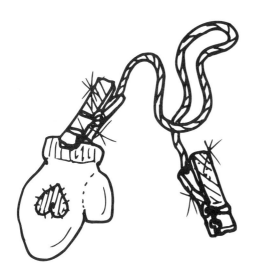

Mitten Dryers

Paint two spring-type clothespins with acrylic paint. Decorate with glitter, if desired. Using strong craft glue or a glue gun (adults only, please), glue a colorful piece of heavy string or yarn to each clothespin. Clip wet mittens to the clothespins and hang over a waterproof surface until the mittens are dry.

Weather Mini-Stick "Puppets"

Reproduce the patterns on page 63 onto heavy paper. Color and cut out. Glue to craft sticks. Decorate with glitter or sequins, as shown. Use the finished "puppets" to illustrate weather lessons and for weather-related plays and skits, using a shoe box as a stage.

Craft Patterns

Weather Fingerplay

Look around and what do I see?
(Children put hand above eyes and look around, side-to-side.)

Lots of weather just for me!
(Proudly point to chest and smile.)

Big, puffy clouds, so high in the sky,
(Put both hands high in the air, stretching body.)

And lots of wind for my kite to fly!
(Wave arms from side-to-side.)

Raindrops fall on my hair,
(Put hands on top of head and tap hair with fingertips.)

Chilly, chilly breeze! I need a coat to wear.
(Hug self and pretend to shiver. Pretend to put on coat.)

But my coat comes off when I see the sun.
(Pretend to take coat off and brush hand over forehead like hot. Make a circle with arms.)

Watching the weather sure is fun!
(Make pretend "binoculars" with hands and look up at sky. Big smile!)

Weather
ACTION RHYME

I Like Weather!

I like hot weather, the hotter the better.
(Wipe brow with back of hand.)

There's nothing so fine as a day of sunshine.
(Look up at the sun with hand shading eyes.)

But I also like rain on my windowpane.
(Wiggle fingers like rain pouring down.)

It splashes below and makes the plants grow.
(Move hand up to indicate growth.)

Storms can be frightening with thunder and lightning,
(Smack hands together and make crashing noises for thunder.)

But they light up the sky like the Fourth of July.
("Flash" hands open like fireworks exploding.)

When cool breezes blow, it's outside I go.
(Make wind noises and sway back and forth.)

I sail my kite high, way up in the sky.
(Pretend to fly a kite.)

The hard-blowing wind makes giant trees bend.
(Stand with outstretched arms, swaying and bending in the wind.)

I like to look out and see leaves blow about.
(Flutter hands like blowing leaves.)

Outside in the snow is a great place to go.
(Twirl around with arms out and face uplifted.)

The snow makes no sound as it falls all around.
(Hold index finger up to pursed lips and say, "Shh.")

I'll be inside and warm when there's an ice storm.
(Hug yourself.)

The ice glitters and shines on trees and clotheslines.
(Shade eyes and squint.)

Snow, rain or sunshine–I think every day's fine!
(Jump up with hands over head and say, "Yes!")

Snacks

Rain and Clouds

Fill clear plastic cups with blueberry-flavored gelatin, leaving one inch at the top. When the gelatin has set, top with a "cloud" of whipped topping.

Rainbow Salad

Have children create a rainbow snack using small pieces of fruit. They can arrange the pieces in layers in a clear bowl or cup.

     purple: grapes
     blue: blueberries
     green: kiwis
     yellow: bananas or pineapple
     orange: oranges
     red: strawberries

Snowman Fun

Stack three large marshmallows together, using white icing as "glue." Add pretzel stick "arms" and draw a face using a toothpick dipped in food coloring. Add a Fruit Roll-Up™ "scarf," if desired.

⚠ Make sure you are aware of any food allergies or restrictions children may have. Be sure children wash hands before they eat.

My
Book
About
Weather

_____
Name

Flowers like rain! Draw three flowers.

1

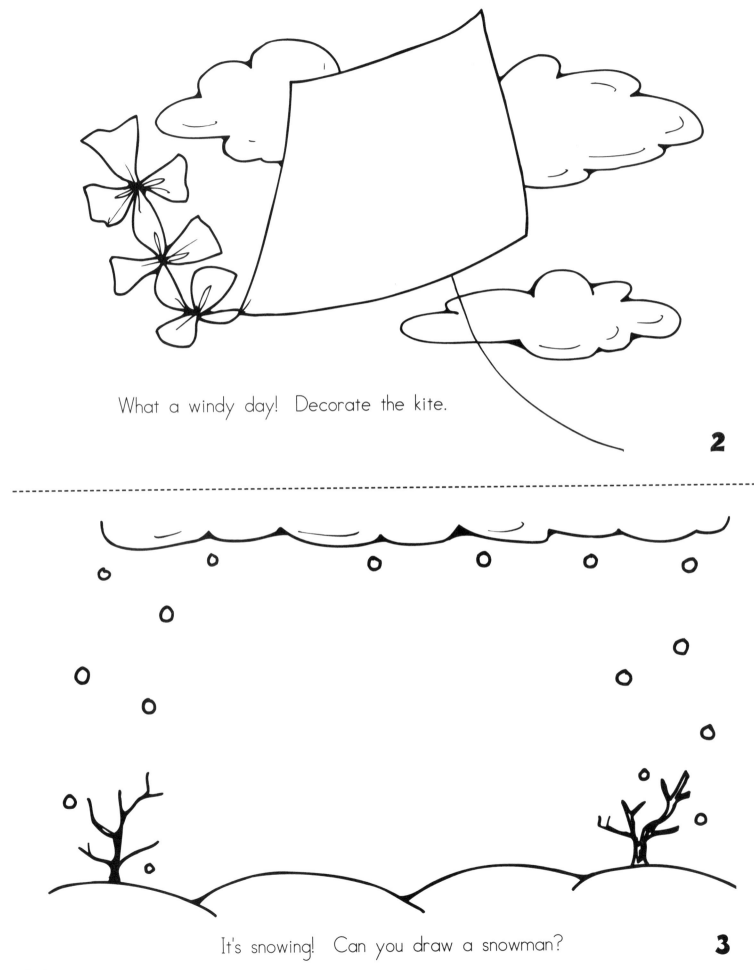

What a windy day! Decorate the kite.

2

It's snowing! Can you draw a snowman?

3

It's hot! Draw a big sun in the sky.

4

Weather is wonderful!

5

Colorful Fun

Color each picture the correct color.

gray

blue

green

pink

orange

yellow

Name

Count and Color

Follow the directions.

Color the third kite green.
Color the sixth kite red.

Color the second kite blue.
Color the fifth kite purple.

Color 2 mittens black.
Color 1 mitten brown.

Color 3 mittens red.

Color the fourth cloud pink.
Color the second cloud yellow.

Color the first cloud blue.
Color the fifth cloud purple.

Color the third raindrop orange.
Color the first raindrop pink.

Color the last raindrop yellow.
Color the fourth raindrop green.

Name _____

Number Skills

Count the objects in each box. Circle the number that shows how many are in each set.

1 2 3 4 5 6 7 8

1 2 3 4 5 6 7 8

1 2 3 4 5 6 7 8

1 2 3 4 5 6 7 8

1 2 3 4 5 6 7 8

1 2 3 4 5 6 7 8

72

Dot-to-Dot

Connect the numbers to see what the weather is like.

Matching Sounds

Look at the pictures. Think of the letter that makes the first sound of each.

Draw a line from the picture to the correct letter.

s

r

k

c

h

Weather Maze

Can you help the cloud find the puddle?

75

Name _____

My Weather Page

Book Cover

All About Weather

_____
Name

Weather Poster to Color